AF270693

SKUNKS

by Elizabeth Andrews

Cody Koala
An Imprint of Pop!
popbooksonline.com

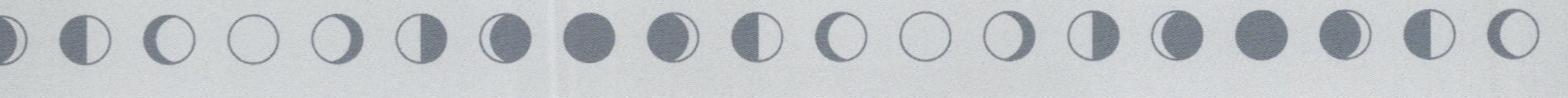

abdobooks.com

Published by Pop!, a division of ABDO, PO Box 398166, Minneapolis, Minnesota 55439. Copyright ©2023 by Abdo Consulting Group, Inc. International copyrights reserved in all countries. No part of this book may be reproduced in any form without written permission from the publisher. Cody Koala™ is a trademark and logo of Pop!.

Printed in the United States of America, North Mankato, Minnesota

052022
092022

THIS BOOK CONTAINS
RECYCLED MATERIALS

Cover Photo: Shutterstock Images
Interior Photos: Shutterstock Images, Christian Sanchez/Getty Images, Bill Gorum/Alamy Stock Photo

Editor: Grace Hansen
Series Designer: Laura Graphenteen

Library of Congress Control Number: 2021951841
Publisher's Cataloging-in-Publication Data
Names: Andrews, Elizabeth, author.
Title: Skunks / by Elizabeth Andrews
Description: Minneapolis, Minnesota : Pop, 2023 | Series: Twilight Animals |
    Includes online resources and index
Identifiers: ISBN 9781098242107 (lib. bdg.) | ISBN 9781098242800 (ebook)
Subjects: LCSH: Skunks--Juvenile literature. | Twilight--Juvenile literature. |
    Nocturnal animals--Juvenile literature. | Nocturnal animals--Behavior-
    -Juvenile literature.
Classification: DDC 591.518--dc23

Hello! My name is

Cody Koala

Pop open this book and you'll find QR codes like this one, loaded with information, so you can learn even more!

Scan this code* and others like it while you read, or visit the website below to make this book pop.

popbooksonline.com/skunks

*Scanning QR codes requires a web-enabled smart device with a QR code reader app and a camera.

Table of Contents

Chapter 1

Neighborhood Skunk

Skunks live in all kinds of places like forests, deserts, cities, mountains, and even in your neighborhood! They live in dens made by other animals and natural **cavities**. Skunks like to live alone.

Watch a video here!

Skunks are the size of
a house cat. They have
long bushy tails and white
markings on black fur. Often
there are two stripes down
their backs. Sometimes they
have spots.

Chapter 2

Stinky Skunks

It's easy to see a skunk as the sun goes down. They like to hunt when night falls. Their black and white markings serve as a warning to possible **predators**.

Learn more here!

Some of the skunk's predators are owls, eagles, coyotes, and bobcats.

Most animals fear skunks because of their **defense strategy**. First, a skunk will stomp its front feet, hiss, and raise its fluffy tail. If the predator doesn't **retreat**, the skunk will spray.

How Far Can a Skunk Spray?

The skunk shoots a bad

smelling liquid from its

back end. This scares the

predator and can sometimes hurt it if the spray gets in its eyes.

Chapter 3

Sleepy Skunks

Some skunks live in places that get very cold in the winter. They go into winter dormancy. This means they sleep deeply for long periods of time. They wake when the weather gets above 20°F (−6°C).

Some skunks share
dens to add more body
heat to their spaces.

Explore links here!

Skunks eat a lot of food during the fall to prepare for winter. This builds a layer of fat that will keep them warm. Skunks are **omnivores** that eat plants, insects, fruit, and rodents.

Chapter 4

Little Stinkers

Female skunks have between two and 12 **kits** per **litter**. Sometimes they share a den with another mother skunk. Male skunks do not help raise the kits.

Complete an
activity here!

When the kits are born,
they stay in the den for six
to eight weeks. After, they

follow their mother around
and learn to hunt for food
and find dens.

Making Connections

Text-to-Self

What would you do if you ran into a skunk?

Text-to-Text

Have you read any other books about skunks? If so, did you learn anything that wasn't covered in this book?

Text-to-World

Do you think skunks deserve the be known as stinky creatures or is there more to them than that? Explain your answer.

Glossary

cavity – a hollow place or hole.

defense – a way of protecting one's self.

kit – the young of some fur-bearing animals such as beavers, raccoons, and skunks.

litter – a group of young animals born to one mother at the same time.

omnivore – an animal that eats both plants and other animals.

predator – an animal that hunts other animals for food.

retreat – the act of moving back or away from a place or situation.

strategy – a careful plan or series of actions.

Index